GLOBAL WARMING

THE CHRISTIAN AND CLIMATE CHANGE

GLOBAL WARMING

THE CHRISTIAN AND CLIMATE CHANGE

Bert Cargill

RITCHIE

John Ritchie Publishing

40 Beansburn, Kilmarnock, Scotland

ISBN-13: 978 1 914273 38 4

Copyright © 2023 by John Ritchie Ltd.
40 Beansburn, Kilmarnock, Scotland

www.ritchiechristianmedia.co.uk

Typeset by John Ritchie Ltd., Kilmarnock
Printed by Bell & Bain Ltd., Glasgow

In *A Christian's Guide* series of books, the ground rules and principles to follow will be found in the Bible. For this one about *Climate Change*, here are some relevant quotations.

In the beginning
God created the heavens and the earth.
(Genesis 1.1)

The earth is the LORD's, and all its fullness.
(Psalm 24.1)

Upholding all things by the word of His power ...
(Hebrews 1.3)

While the earth remains,
seedtime and harvest, cold and heat,
winter and summer, and day and night
shall not cease.
(Genesis 8.22)

We, according to His promise,
look for new heavens and a new earth,
in which righteousness dwells.
(2 Peter 3.13)

This same Jesus ...
will so come in like manner
as you saw Him go into heaven.
(Acts 1.11)

(All quotations from New King James Version of the Bible)

CONTENTS

PREFACE

This matter of climate change and global warming has recently become a very hot(!) topic in the news and elsewhere. Over the last few decades, more people than ever have taken a keen interest in it, even have a real passionate concern about it and its effects. Of course there are a few 'deniers' who are suspicious and doubtful about it, while others find it debatable and controversial. And a few don't seem to notice! I hope you will find it interesting enough to give it some thought, and that what you read here may help you to see it from a Christian's point of view.

Before we get down to it, it is essential to remember that the Christian viewpoint on this can be quite different from that of other people - as is the case in many other matters also. Almost all we hear and read about climate change and global warming, and how to interpret the science behind it, comes from a world view which leaves God out of the picture altogether. It is a materialistic viewpoint which regards the past, the present and the future as solely under the influence of uniform natural forces. No room is allowed for divine revelation or supernatural input and intervention.

The Bible tells us that God has a definite plan for the future of planet Earth and everything else. We are assured that our Lord Jesus Christ is coming "a second time"[1] and huge changes will

---

1 Hebrews 9.28

begin to happen, with great blessing for some people and awful disaster for others. And Christians know that "the word of the LORD endures forever";[2] and that "the fear of the LORD is the beginning of wisdom".[3] These great principles will govern our treatment of this subject, so please read on!

Bert Cargill
St Monans
Scotland
2023

---

2 1 Peter 1.25
3 Proverbs 9.10

CHAPTER 1
THE WEATHER, THE CLIMATE AND THE CHANGES

What is the weather like?

I'm writing this during a wet, grey November day on the Scottish east coast with high waves from the North Sea driven by an easterly gale crashing over the harbour wall nearby. Rain is forecast for the rest of the day, and I'm thankful to be warm and dry inside. For travellers on land, on the sea and in the air it is much less pleasant, and for some it is downright dangerous. For workers outside, well-designed clothing enables them to continue their tasks with some degree of warmth and protection. For others with cold and damp housing, or no shelter at all, their misery is compounded, and for them the outlook is bleak. Many of us have cause to "count our blessings" while many others are having to count their expenses and consider what to do next. And I'm thinking only about ourselves in this country just now, and I'm writing from a rather privileged perspective. In other countries severe weather events have been catastrophic and even fatal. There the perspective will be very different.

We've had this kind of weather many times before, and worse, and we'll likely get it again. In the UK we get a fair balance of what we call good weather and bad weather. We pay attention to weather forecasts and hope they are right when we make our plans, but we are accustomed to it changing. It's even a talking point – "as changeable as the weather" you may have heard people say! But our local weather pattern doesn't seem to have changed dramatically for as long as we can remember, or has it? Maybe it depends on your selective memories as time passes! But elsewhere there have been big changes in weather patterns, and some people in other countries are suffering badly because of it.

What is the climate like?

Weather is a local and temporary thing, and it has always changed to and fro between certain limits of temperature, wind speed, rainfall and so on. But climate is something different - it is the long-term averaged meteorological characteristic of a large area. Over the planet it has traditionally been classified broadly as arctic, temperate, and tropical, each with its own characteristics, and with some subsets such as maritime and mountainous.

But is the world's climate really changing as we are being told repeatedly nowadays? Is there global warming? What is the evidence? In the depth of winter during an exceptionally cold spell, we may be tempted to think that global warming isn't real, or even think that it could be an advantage to us!

Over the whole of planet Earth, averaged meteorological measurements over short and long periods are showing definite changes and trends. The climate does appear to be changing, not in a random fashion to and fro, but steadily in one direction. Measurements are showing that the world is getting warmer year on year. 2022 looks like being the warmest year on record so far. The big concern is what this will mean for everyone everywhere and for future generations if the trend continues as it threatens to do. The big question is: What can be done about it? Can this problem be solved? Can the trend be slowed down or even reversed?

What is the Christian viewpoint?

What should a Christian's attitude be to this concern over climate change, to what we are told is happening now, and also what is expected in the near and distant future? It is easy for us to be a bit selfish and think locally and short term, and believe that it doesn't affect us much here and now. The effects may indeed be slight and of short duration for most of us in the UK and in similar countries at present. But that is not the case in many other parts of the world, especially for poorer, underdeveloped countries. Many of them have already, and for a long time, had to cope with the terrible effects of prolonged droughts, of severe floods, of uncontrolled fires, of fierce hurricanes and the like. Such weather/climate phenomena have already wreaked havoc on lives and livelihoods in many places worldwide. Global warming is predicted to make it much worse. Christians cannot afford to "look the other way" when confronted with human suffering.

We will look at all this as carefully and sensitively as we can. But first we need to try to analyse the whole problem in some way, to understand why we are where we are now in respect of the world's climate. So in the next few chapters we will look into the following areas:

- The evidences of climate change and the effects of global warming.
- The scientific background, how the evidence is interpreted in keeping with established chemical and physical principles.
- The historical background, how we have arrived where we are now.

- The efforts which are being made and proposed currently, worldwide, to mitigate the problem.

- Some of the difficulties being faced in reaching international agreements.

- What we as individuals can do about it.

You can skim over any of these areas that you find too complex (the subject is actually very complex!), and get to some conclusions which you feel are appropriate for you as a Christian to believe and act upon at this time.

What changes should we expect?

As Christians, we know that God is in control of the world's history and its destiny. God Himself and His ways are "past finding out"[4] to human reasoning, but His revelation is real and reliable. So we must keep coming back to what the Bible tells us about the future of planet Earth and everything else. We are told that God will intervene and bring about change, dramatic and far-reaching change, in His own time and in His own way. We have been told about this in the many prophetic scriptures in the Bible, a subject dealt with more fully in other publications.[5] A brief summary is as follows.

We know that the Creator, our Lord Jesus Christ, is also the upholder of all things by the word of His power (Hebrews 1.2-3). We know that He is coming again according to His promise,[6] and that when He comes He will set in motion a programme for deep and lasting change. First He will take from earth to heaven

---

4 Romans 11.33
5 *Tell me more about the Future*, Bert Cargill, 2017 (a short one); *End Times for Beginners,* M David McKillen, 2021, (a lengthy one) - both John Ritchie Ltd.
6 John 14.1-3

> **These great changes will occur in God's time and in His way 'according to the counsel of His will' (Ephesians 1.11)."**

every true believer whether alive at the time or resurrected from where they may have lain for centuries. We often call this the Rapture – details in 1 Thessalonians 4.13-18.

Then on earth there will be a period of turmoil, suffering and destruction on an unprecedented scale affecting the world's population and the whole environment, central to which is the nation of Israel. The Bible calls this "the great tribulation", described in some detail by our Lord Jesus Himself in Matthew 24. It will conclude with the appearance of Christ Himself as warrior King of kings and Lord of Lords, defeating every enemy of God and man, then inaugurating His kingdom on earth which will last for 1,000 years (the Millennium), characterised by righteousness, peace, and joy. You can read about this in Revelation 19-20.

These great changes will occur in God's time and in His way "according to the counsel of His will" (Ephesians 1.11). There is actually a beautiful future for planet Earth![7]

Multiple Concerns

Before we leave this chapter it is worth noting that this concern about climate change is only one of many potential crises facing and alarming the present generation. There are multiple concerns about the pollution of the air we breathe, and the pollution of much of our land and of our seas, particularly

---

7 Read for example Isaiah 2.2-5; 11.1-10; 35.1-10.

by plastic; about the exhaustion of finite mineral resources, particularly some metals which seem essential to our lifestyles; about the destabilising of civilisation by wars and terrorism; about pandemics and new diseases which antibiotics and other drugs cannot yet deal with; also about the proliferation of narcotic and similar harmful drugs (and of armaments including nuclear ones); about extremes of poverty and the lack of public funding for essential infrastructures; about the polarisation of racial groups; and about overpopulation and migration of refugees. Earthquakes, volcanic eruptions and tsunamis add to the problems, as do the instabilities of economic and financial systems worldwide, and the erosion of traditionally dependable moral standards of behaviour in business life and in governments as well as in personal life.

Indeed to many, the world seems to be descending into chaos. Could someone not take worldwide control over all the threatening autocratic governments in the world, and over the diverse liberal democratic ones, each with its own selfish agenda? Later we'll find that such a scenario is described in the Bible![8]

However, as Christians nowadays think about all these concerns over the environment and about all these crises in world systems, they have cause to believe that their coincidence, and their increasing intensity, point to the fact that the second coming of the Lord Jesus Christ must be near at hand, bringing to an end this period of God's grace to mankind. Up till now, and for many long centuries, God has been longsuffering, "not willing that any should perish but that all should come to repentance. But the day of the Lord will come ..."[9]

---

8 For now you could read 2 Thessalonians 2.1-17.
9 2 Peter 3.9-10

To think more about

Do you think that natural disasters, man-made miseries, and political instability are worse nowadays than ever they have been?

As we become aware of these crises all around us, should this not motivate us even more, to pray and to mobilise our efforts to reach the present generation, young and old, with the gospel of salvation through repentance and faith in Christ before it is too late?

GLOBAL WARMING

CHAPTER 2
EVIDENCE OF CLIMATE CHANGE
AND GLOBAL WARMING

Since the latter part of the 20th century, changes in the severity of weather events in many parts of the world have become more and more noticeable. Each passing year brings more to our attention - devastating floods of increasing extent, heat waves of greater intensity, terrible wild-fires spreading out of control, more tropical storms, glaciers receding more and Arctic ice melting faster, hotter summers and milder winters even in the UK, temperature and rainfall records broken more frequently, and, as a consequence, more people suffering and dying. It is an alarming picture, said to be a direct result of global warming.

Take, for example, the year 2022, recently noted as the warmest year on record. Extended months of drought in some northern African countries like Somalia have caused awful famine and loss of life. On the other hand, some southern African countries like Malawi suffered from serious flooding. Pakistan had it worse. Annual floods due to the Monsoon are expected there, but, on this most recent occasion, really terrible floods from the River Indus devastated huge areas, uprooting thousands of families from their homes and allowed water-borne diseases to spread. Almost 2,000 people were killed and 33 million more were affected. The United Nations called it the worst humanitarian crisis in the world. (In 2010, the death toll was similar, and 21 million people were affected.)

... there are many changes which almost anyone can observe."

Perhaps more surprisingly, floods also affected some more developed countries such as Australia, parts of which also experienced extensive wild-fires. And some flooding and forest fires have also been more common than ever, and more extensive, in parts of Europe and the USA. In the UK, temperature records and rainfall records have been broken for different months of the year. The list of catastrophes goes on – coastal erosion, typhoons and hurricanes – all combine to make life more difficult and the future more uncertain.

In addition to these extreme events, there are many changes which almost anyone can observe. In recent years there have been signs of a shift in the seasons, spring arriving earlier and winters tending to be less severe at times. These seasonal changes affect the nesting times of birds and the migration patterns of birds, insects and mammals. Plants are also blooming earlier in our gardens. Some fish and other marine species are moving farther north around our coasts.

Photographic evidence and site visits over the past 100 years or more have shown that glaciers in mountain ranges in many parts of the world are melting and receding to a fraction of their original size. In the Alps, for example, until about 1900, France's largest glacier, the *Mer de Glace* near Chamonix, filled the valley up to the level of the railway station built in 1908 to enable tourists to visit it. In fact, they could step out of the train and on to the glacier right there. Now you have to descend 580 steps to reach its surface - 100 metres lower down, down past markers on the rocks showing where the surface of the glacier was just a few decades ago. It has been melting and shrinking measurably faster ever since 1990.

Snow cover is also declining in general in the Northern Hemisphere. Greenland's ice sheet (it holds about 8% of Earth's fresh water) is melting faster. The area and the thickness of Arctic sea ice is reducing rapidly with large icebergs breaking off into the sea, and the effect of this on its wildlife has been well publicised.

Global sea level is rising gradually due to this melting and the general temperature increase all over. Measurements show that between 1900 and 2019, it rose by 21 centimetres (8 inches). If this continues, many low lying coastal communities will be severely threatened.

The common factor behind all these phenomena is a definite increase in Earth's temperature which can now be carefully measured. Since the year 1900, average surface air temperature has increased by around 1°C. The period between 2010 and 2020 was the warmest decade on record, with 2016, 2020, then 2022 being the warmest years since records began. The changes may appear small, but they are significant. As we shall see, such changes are magnified into bigger ones in the interlinked environmental systems. The scientific consensus just now is that further temperature increases must be kept below 1.5°C if catastrophic and runaway global warming is to be avoided with disastrous results for all life on earth. No wonder some people and organised groups are protesting and clamouring for governments to take more immediate action.

The Basic Picture

The effects of global warming are many and varied - we have described only some of them and noted that they have been disastrous for many people across the globe. But the overall reason is not difficult to describe in basic terms. You could work this out from a very basic knowledge of science, or from everyday observations in your kitchen as you cook or boil a kettle!

As the global temperature rises even by an apparently small amount, here is what happens all over the world.

- More water evaporates from the surface of the earth in the form of moisture or water vapour, so there is more dried-out vegetation, crop failures, and wild-fires. There is also more local drought as this moisture is carried to other areas.

- The extra water vapour now in the atmosphere accumulates until the clouds become saturated and water precipitates as extra heavier rainfall. Clouds are mobile and driven with winds so that the heavy rainfall and consequent flooding can occur anywhere, but especially in places prone to heavy rainfall and flooding due to the geography of the land.

- Extra water vapour in the atmosphere traps some more heat and retains it, but it is extra carbon dioxide in the atmosphere which is the main driver of global warming and consequent climate change.

To think more about

These are just some of the evidences for global warming which is leading to climate change. Have you noticed any of them yourself, or other ones?

Have they affected you personally in any way?

In the next chapter we will try to understand why all this is happening, and why the main culprit is a gas called carbon dioxide which we cannot avoid.

CHAPTER 3
THE SCIENCE BIT

Temperatures on Earth

God created planet Earth to be a suitable place for man to live in, and for the great work of redemption to take place for the glory of God and the eternal blessing of man. Conditions on Earth were designed to support life, including the whole variety of plant life and animal life on which we depend. For this, the temperature on Earth is crucial, particularly to keep water in liquid form. The worldwide average is about 15°C, although localised extremes can be over 50°C and under -50°C.

The Goldilocks Effect

You remember the children's story about the naughty little girl called Goldilocks (do you remember why she got that name?). In her bold explorations of an empty cottage in the woods she eventually found porridge that was "not too hot and not too cold", it was "just right"! Well, this name has been given to the combined factors which make the temperature on Earth "just right".[10] It is principally because planet Earth is just the correct distance from the sun, 93 million miles away. Farther away from the sun we would freeze to death; nearer it we would be overcooked, if you will allow these terms to be used! Materialists make the outrageous claim that this just happened by chance

---

10 Paul Davies, *The Goldilocks Enigma,* 2007.

millions of centuries ago, but we know that Almighty God in His wisdom planned it this way, as He did for every other fine-tuned process, from the galaxies of the universe right down to the fine details of the smallest living cell and the tiniest atom which each contains.

So we begin by noting that Earth's temperature is first of all controlled by the Sun, these 93 million miles away. It radiates heat and light constantly into space in all directions and Earth receives its share 8 minutes after it leaves the sun. Now about 30% of this incoming sunlight is reflected back into space by bright surfaces like clouds in the sky and snow and ice on the ground. Of the remaining 70%, most is absorbed by the land and the oceans, and some by the atmosphere. It is the absorbed solar energy which heats our planet.

The Greenhouse Effect

Here is another more commonly known term, used this time to describe the effect of the earth's atmosphere on its temperature. Basically the atmosphere is likened to the glass of an everyday greenhouse which is there to retain the heat inside, to insulate it.

As the land, the air, and the seas warm up, they radiate heat energy outwards into space (thermal infrared radiation). As this energy goes up into the atmosphere, much of it is absorbed by the water vapour which exists in large amounts, and by gases such as carbon dioxide and methane which are there in much smaller amounts, but are much more effective at absorbing heat. They are called the "greenhouse gases".

When these gases have absorbed the energy radiating out from Earth's surface, their molecules radiate some of that heat back towards Earth. It adds extra heat to both the lower atmosphere and the earth's surface, enhancing the warming effect produced by the direct sunlight. It creates an overall insulating effect. A very readable account of all this with a helpful diagram can be found at https://climate.nasa.gov/causes *Causes of Climate Change.*

Before examining the problems associated with greenhouse gases, it should be noted that this insulating effect by the atmosphere is actually beneficial, indeed necessary, for life on Earth. If there were no such natural greenhouse effect, Earth's average surface temperature would be much lower, perhaps *minus* 18°C, instead of the plus 15°C that we have today. Our atmosphere serves many useful and beneficial purposes simultaneously, in addition to the obvious one we take for granted, providing oxygen for us at the "just right" concentration for us to breathe.

Greenhouse Gases

Water vapour itself is a greenhouse gas by definition, because it retains heat radiating from Earth's surface. The amount of it in the atmosphere depends directly on the temperature. The warmer the earth, the more water evaporates into the atmosphere, but rainfall in the water cycle creates a balance. So the amount is stable enough to be disregarded as a cause for concern. Methane is a more potent greenhouse gas, but this time the amount is quite small (but increasing) and it disappears fairly quickly. It is produced naturally from the decay of vegetation and from farming livestock, as well as leaks from fuel production. Our natural fuel gas is around 90% methane.

The gas in the atmosphere which has the biggest greenhouse effect to concern mankind is carbon dioxide, CO_2. It is more stable than methane and there is more of it, in fact more of it than ever, and that brings us to the current problem. It is produced

Up until about 200 years ago, the amount of CO_2 in the atmosphere was fairly steady."

naturally in volcanic eruptions and forest fires etc. Along with all animal life, we continuously breathe out carbon dioxide to add to the total, but it is our industrial activities which have greatly increased the amount in the atmosphere. Up until about 200 years ago, the amount of CO_2 in the atmosphere was fairly steady, just below 300 ppm,[11] but since then it has increased by around 40% to 420 ppm today, and it is rising steadily. In the year 2000 it was 370 ppm, increasing by 11% in just 22 years to what it is now.

When fuels like coal, oil, gas, and wood are burned to produce heat, the carbon they contain combines with oxygen from the air and changes into carbon dioxide, CO_2. The chemistry is quite simple and you could maybe write a one-line equation to describe it. If you did, the products would be carbon dioxide and probably water as well from the hydrogen content of the fuel. To make the equation complete, however, you would need to include heat energy as a major product. This is, of course, the main reason why combustion of fuels is so important to us, as it has been to mankind from earliest times.

Nowadays more industrialisation requires more energy, more travel uses more fuel, more homes need to be heated and more food needs to be cooked. All these processes produce huge amounts of CO_2 and are accompanied by excess heat being inevitably released into the environment. So you can see why the amount of CO_2 in the atmosphere has increased so much over the past 200 years, and how all this has been linked to global warming.

---

11 ppm is 'parts per million'. 300ppm = 0.03%.

We have noted already that measurements over the past 100 years or so indicate that the average temperature of the earth's surface has increased by over 1°C. This may seem small but it is significant amount. And it is still going up as more and more CO_2 is being added to the atmosphere. What can be done to control this? Could it be stopped or even reversed? How?

Read on into the next chapter.

To think more about

The greenhouse effect is essential to maintain life on Earth, but man-made emissions into the atmosphere are exaggerating it as they reduce and slow down the loss of heat into space. This extra retention of heat is what is causing the warming on the planet's surface and changing the climate.

Each of us, directly and indirectly, and to a greater or lesser extent, is contributing to it!

CHAPTER 4

THE CHALLENGE OF CO₂ IN THE ATMOSPHERE

A Balanced Planet

There is a remarkable balance in the living world between processes which create CO_2 and those which remove it. Generally speaking, the animal kingdom produces and emits CO_2 and uses up oxygen from the air for breathing, while the vegetable kingdom absorbs CO_2 by photosynthesis, converting it into biomass, at the same time producing and emitting oxygen.

This occurs in the sea as well as on land because the large amount of vegetation in the sea (which we usually call "seaweed") grows like other plants do by photosynthesis. Also there are large amounts of phytoplankton in the sea which actively grow by the same photosynthetic process. Driven by sunlight filtering through the surface, all this again converts carbon dioxide into the fabric and tissues of the growing plankton and seaweed and produces large amounts of oxygen. In addition to all this, the water itself dissolves large amounts of CO_2. In fact, the sea is responsible for removing just about the same amount of CO_2 as all the forests and vegetation on dry land, and releases just as much oxygen.

If this balance between creating and removing CO_2 was maintained, there would not be any problem of global warming. But there is a problem because, on the one hand, burning more

No wonder there is a problem - more into the atmosphere and less removed!"

and more carbon based fuels has overloaded the atmosphere with CO_2, and on the other hand, at the same time, large swathes of forests have been felled in many countries, and greenery is always being reduced through clearing for industry and buildings. So the amount of CO_2 which is absorbed is decreasing while the amount being produced is increasing. No wonder there is a problem – more put into the atmosphere and less removed! This situation has been getting worse for years now.

In addition to all this, there is another factor to consider. The solubility of a gas in a liquid decreases as the temperature increases, so as water gets warmer it can dissolve less of any gas. So the warming of the water in the seas (and in other surface water bodies) makes it less able to dissolve CO_2. Along with this heating effect on solubility, we know that water expands on heating. This will add to the problem of rising sea levels, extra to what is being caused by the continual melting of sea ice into the sea at the poles We must not overlook the importance of the sea in controlling global warming.

Here are two examples of how some factors can combine and reinforce each other to make the global warming problem worse.

(1) Warmer sea water will dissolve less CO_2, leaving more in the atmosphere which will increase the greenhouse effect and so raise the temperature more.

(2) Less ice and snow due to global warming means that less heat is reflected back into space from these bright surfaces on Earth, and so its temperature will increase more.

These are examples of what is sometimes called a positive feedback loop.

"Saving the Planet"

These facts and the potential threats to the future of the planet are now well recognised, backed up by much detailed science and by many ordinary observations. The future wellbeing of the world's population and to almost all of its wildlife is at stake. You might well ask: What is being done to tackle the problem? Indeed, what can be done to "Save the Planet", as is the slogan?

Clearly there are two ways to go about it and more and more attention is being given to each of them now. Both of these are necessary, but none are without difficulties. Some of the difficulties are practical and technological, requiring more research and development. All have massive financial implications requiring huge expense and investment, and also claims are being made for compensation for damage deemed to have been caused already. Not least are the political factors within and between different countries as each strives to protect at least its own interests, its own industries and economies, according to its own ideology.

The two broad approaches to reducing global warming are as follows, both of them looking promising on paper.

1. Produce less CO_2 so that less would be released into the atmosphere worldwide.

    This means reducing, and as soon as possible halting, combustion of all carbon based fuels, reversing the trend of the past 200 years and more.

2. Increase, protect, and add to mechanisms for removing CO_2 from the atmosphere.

This means stopping large scale deforestation as soon as possible, planting more and more trees, and protecting green spaces, again reversing the trend of perhaps 2,000 years! Also apply science to find and develop new technologies for "carbon capture", somehow collecting the CO_2 at its source before it is released into the atmosphere.

More about all this in Chapters 5 and 6.

To think more about

Should Christians be concerned about saving the planet, or more concerned about saving precious souls?

Can it be both? And what does "saving" really mean here?

CHAPTER 5
DEALING WITH THE CO_2 PROBLEM: (1) REDUCING CO_2 PRODUCTION

Dealing with the CO_2 problem means first reducing its production somehow. So instead of producing energy by burning carbon based fuels, different sources of energy will be needed. Let's survey the traditional sources of energy and notice how they have changed already, then look at the alternatives.

Fuels

Historically it was **wood** that fuelled the fires of our ancestors for primitive heating and cooking, as is often mentioned in the Bible, and as is still commonly done in many rural settings. Charcoal (partly charred and totally dewatered wood) is the fuel of choice for cooking and heating in several rural communities in many African countries. Nearer home, there has been a resurgence of using wood or "biomass" as fuel in domestic stoves, and in some industrial installations to generate power and supply heat for buildings. This has been seen a positive move towards the use of "renewables", but it does not help the big objective of reducing the production of CO_2. In fact, burning wood is simply releasing again into the atmosphere the CO_2 which was absorbed from the atmosphere by the tree or plant when it was growing years before.

Then **coal** was found in some places and at first was quarried manually from outcrops and fairly shallow mines in the ground.

> # Carbon dioxide is the inevitable product of all carbon based fuel combustion, and ... is the chief culprit in the global warming scenario."

This soon developed into a major industry with deep pits being sunk to access rich seams of prime quality coal underground, now extracted by machinery in huge tonnages. In the UK it was often called "King Coal" and for over two centuries it provided the heat to power steam engines and support the Industrial Revolution. Many other countries soon caught up and during the 20th century, coal was the dominant fuel for power stations to generate electricity.

The discovery of **oil**, and developing the technology for refining it to get diesel and petrol, led to the steam engine being replaced by the more efficient internal combustion engine, especially for transport. It has also been replacing coal for domestic and industrial heating, and for generation of electricity. Then large underground reservoirs of natural **gas** (methane) were discovered in many locations, and technology quickly developed to deliver it to users through hundreds of miles of pipelines, underwater and on land. It all became a huge and profitable industry. One result of this was that gas was soon replacing oil, for space heating, cooking, and for some power generation too.

All these changing and developing sources of energy created benefits for consumers and for many industries, and huge financial profits were made by the organisations which did the hard work of exploration, production and supply. At the same time, governments benefitted from large taxation revenues. Now all this is due to change, and change more dramatically than ever! This is because carbon dioxide is the inevitable product of all carbon based fuel combustion, and carbon dioxide is the chief culprit in the global warming scenario.

Alternatives to Fossil Fuels

Electricity is well recognised as being a convenient, versatile and adaptable form of energy. In fact, we are now being told that "the future is electric"! The big change for the future is that electricity will not be generated from the heat given off during fuel combustion, but from different sources altogether, such as the following which do not produce CO_2.

Turbines for electricity generation can be driven (turned at high speed) by running **water**, and also by **wind**, instead of by steam produced by heat from furnaces, and more efficiently too. Indeed, for many years already, large hydroelectric schemes have been built in many countries and used to generate electricity, and a more recent but fast spreading development is harnessing wind power in clusters of wind turbines, so called "wind farms" on land and sea.

Another efficient and rapidly developing method for generating electricity uses **solar panels** which are collections of connected photocells facing the sun. Expensive to begin with, but now more common, these cells absorb sunlight which energises electrons to create a cumulative voltage which can be matched to a normal supply system.

All these alternatives are called **renewables** because they are continuously available and renewed from ongoing processes on Earth and do not use up finite resources which have to be mined or extracted.

It is worth noting that all these renewables are using energy supplied continuously by the sun, not only in solar panels directly but in all of the others indirectly.

For hydro power, it is the sun's heat which first raises vapour from surface water into clouds in the atmosphere, for rain to fall on high ground then gravity to send it downhill, some of it running through a turbine.

For wind power, it is the heat of the sun that causes uneven air pressures in the atmosphere so that wind blows from highs to lows and as it does so it can turn turbine blades. (This is what it did in the old windmills everywhere, and we mustn't forget that wind power enabled thousands of sailing ships to cross the oceans for centuries. The sail was an early form of turbine blade, adapted now in modern racing yachts.)

For wood and all forms of biomass, energy taken from the sun made them grow (as happened in the antediluvian world which provided coal and oil underground following the turmoil of the great Flood[12]).

Renewables are the best choice for future sources of electrical energy - versatile, safe, and 'clean', that is without the production of greenhouse gases. However, for transport applications on road, rail, air, and sea, portable storage of electricity is a challenge. Batteries are the traditional answer, but historically they have been heavy and bulky. Lighter batteries are now being developed and manufactured, essential in the production and promotion of motor vehicles which will be powered by electric motors drawing current from rechargeable batteries onboard, not by using carbon based fuels such as petrol and diesel. In this way, pollution on our roads and cities will be reduced and eventually removed - "electric cars" are to be the future of motoring.

Alongside of this is the development of **hydrogen** as a fuel for vehicles, for it can be stored onboard in specially designed tanks, giving an alternative to the battery problem. It can be burned in a modified internal combustion engine producing (clean) water,

---

12 See Ch 8, pg 54 for more details.

not polluting carbon dioxide. Hydrogen can be produced from water in the first place by passing an electric current through it, a well known process called electrolysis, with oxygen as the other product. To be compatible with the aim of reducing emissions, however, the electricity would need to be generated from renewable sources, not from fuel combustion. Hydrogen is already being produced in this way and used as fuel in some forms of public transport with the objective of reducing urban pollution.

One other source of energy must be mentioned, especially for electricity generation, and it is **nuclear power**. Huge and expensive nuclear power stations have been built in many countries and a considerable proportion of the electricity used in these countries is generated in this way. Heat is produced by nuclear fission - atoms of nuclear fuel such as uranium are split into smaller atoms. When this happens, small amounts of fuel are converted into huge amounts of heat, which again is used to generate steam to drive turbines. However, the by-products of the nuclear reaction are radioactive elements which are toxic. Safe and secure storage of these is a long term problem, and this makes them undesirable to many people. There is no pollution by CO_2 but instead hazardous nuclear waste has to be dealt with, stored securely somewhere for a very long time.

Along with all these proposals and demands for action, it has to be realised that to shut down suddenly all combustion processes which generate energy is not possible. To listen to some climate change protesters, this is what they seem to want! Fossil fuels cannot just be banned overnight. New methods of power generation need to be up and running before traditional sources of energy are abandoned altogether. Also to import such fuels from overseas instead of using local sources is really exacerbating the problem - extra transport costs and extra emissions are involved in this.

Waste

At another level, there is an excessive wastage of energy in present day society which surely needs to be addressed. Take motor vehicles, for example. For general and leisure purposes, is it necessary to have larger and more expensive prestige motor cars, driven at least half empty most of the time? To manufacture and to drive these, more energy is used and wasted, and more pollution is produced than from smaller ones. But more of them seem to be coming off the production lines all the time to meet the demand, including for government and official business (setting a good example?). And what about air travel? – and private jets?

Also in heating most of our older buildings, energy wastage is high. More and better insulation would greatly reduce this, and if the energy saved came from combustion it would obviously reduce CO_2 emissions. In newer buildings, much better insulation is now the norm, and localised energy sources such as solar panels and heat pumps are being used much more, but high initial costs can be a hindrance.

Then at a national and international level, just think with horror about the terrible waste of energy and other precious resources in wars and conflicts which, alas, are still all too common. More CO_2 is produced in this way than is reduced through building hundreds of wind farms. Construction uses energy purposefully, but destruction uses much more energy wantonly in these awful conflicts, not to mention the deeper cost of human suffering and death. Think too about the energy and resources used in development and maintenance of weaponry for defence, and think about all that is used by some countries in prestigious space exploration and development of rockets. None of this comes cheap in terms of finance and energy usage too!

Excessive waste of energy ... surely needs to be addressed."

Christians can see what is behind all this. The woes and troubles of individuals and nations can be traced back to what the Bible tells us is wrong in society. In nations and in each of us, pride and selfishness easily takes over, the fruit of original sin, the springboard for so much trouble and malice. It is worth reading again Romans 3 verses 10 to 18, concluding with, "The way of peace they have not known. There is no fear of God before their eyes".

It will all be different when the Prince of Peace becomes the King who will reign in righteousness. Swords will be beaten into plowshares.[13] Waste and want will be no more – for 1,000 years on this earth!

To think more about

Every source of energy has problems with waste. What would you consider to be the best source, first, if cost was not a problem, and second, taking into account the pollution/waste factors?

Sometimes waste is unavoidable, but most of the time it can be minimised. Notice our Lord's attitude to waste, John 6.12; compare also what He said, Luke 15.13, 16.1; and contrast Matthew 26.8.

---

13 Isaiah 2.1-4; 32.1

CHAPTER 6

DEALING WITH THE CO_2 PROBLEM: (2) MAXIMISING CO_2 REMOVAL

It is easy to add CO_2 to the atmosphere – it happens all the time as we breathe and when any fire burns! Removing it from the atmosphere is more difficult by technological means, although, as we have noted already, it does happen naturally by photosynthesis, that marvellous process which makes all plant life flourish and grow, releasing oxygen at the same time. The more you think about this, the more you examine the details, the more you have cause to marvel at it, and to acknowledge the wisdom and skill which planned and created such an efficient process driven by the light of the sun. Would you say with the psalmist, "O LORD, how manifold are Thy works! In wisdom hast Thou made them all: the earth is full of Thy riches. So is this great and wide sea ..."?[14]

So while technological methods for removing CO_2 are being explored and developed (more about this shortly), the best way forward is surely to use more extensively the natural process which has done it all along. We have noted already that the sea is very effective at dissolving CO_2 and removing it from the atmosphere, although as it warms up it will be less able to do it. We cannot do much about this beyond conservation, but more attention could be given to absorbing and capturing the carbon

14 Psalm 104.24-25 (KJV)

dioxide into growing vegetation where the carbon can stay for a long time – until someone wants to burn it or the organism dies.

Expanding Forestation

Bigger forests and green spaces are desirable, even necessary, and are fine in theory. But, in fact, human activity is steadily reducing forest cover and removing vegetation large and small to enable our projects to proceed - felling trees, excavating peat, replacing green fields by concrete and tarmac, etc. So now the call is to stop or to slow down these processes - in fact, to be planting as many more trees as we can, anywhere and everywhere; also to halt the destruction of large native forests especially in Africa, South America and the Far East where deforestation has been particularly widespread, both legally and illegally. This means reversing the trend of perhaps 2,000 years, because almost the first thing our ancestors did when they moved into a new area was to "clear the land" for agriculture.

The need to preserve and increase green spaces is now being recognised for the sake of a healthy and diverse ecosystem, as well as for carbon capture. For again, humanity's track record in this area is not a good one. Not only has the amount of healthy greenery been drastically reduced, even plundered - under the sea as well as on land – but also the effect of this has been to remove large swathes of natural habitat for wildlife. Because of this, natural populations of many species have declined, with a 69% reduction since 1970 being reported recently. It is also well known that more and more species of both flora and fauna are becoming extinct. As time passes, biodiversity is thus being reduced. A recent conference, COP15, has made resolutions to remedy this by proposing to allocate 30% of all land to natural habitats, not to human activities, by 2030.[15] This

15 United Nations Biodiversity Conference, Montreal, December 2022

The need to preserve and increase green spaces is now being recognised."

would coincide with and reinforce efforts to remove more CO_2 from the atmosphere, but whether it all works out remains to be seen. To limit global warming there is a need to increase, protect, and add to mechanisms for removing the CO_2 from the atmosphere wherever possible.

"Carbon Capture"

Over recent decades, attention has been given to technological methods of "carbon capture", somehow collecting the CO_2 at its source before it is released into the atmosphere. The science and technology for this is already in place, but, as usual, costs are a barrier to large scale application. Money is needed to build and maintain them, and energy is needed to operate them. In outline, the process is as follows.

Flue gases from combustion contain CO_2 in a dilute mixture with other atmospheric gases such as nitrogen. They are passed through a "scrubber tower" to extract the CO_2. In this tower a water-based mixture of a chemical substance[16] attracts and binds the CO_2 only, then this mixture is led into another vessel where at a somewhat higher temperature the CO_2 is released for collection in a concentrated form. The absorbent can be used over and over again. The collected CO_2 could be collected and used commercially, e.g., for soft drinks, animal welfare, or solidified into "dry ice" which is widely used as a low temperature refrigerant, etc. More usually, though, the main purpose of all this is "carbon capture"- it is piped in large

---

16 It is called monoethanolamine

amounts to suitable underground or undersea reservoirs from which oil and gas have already been extracted. At the high pressures deep down underground, the CO_2 is in such a state that it is not likely to escape, and it may even be solidified into some rock formations. Some of these installations are already in place, e.g., in USA where coal is still widely used as fuel for power generation, and this is said to be giving some more time to phase out the use of these fossil fuels.

To think more about

Which method(s) of reducing and removing carbon dioxide do you believe holds the most promise just now?

CHAPTER 7
INDIVIDUAL, NATIONAL AND INTERNATIONAL EFFORTS

On planet Earth there are now around 8 billion people. We all share Earth's resources, breathe its atmosphere, and leave our mark on it. Our time on earth is short, creatures of time but bound for eternity some day. And it is certainly more important to be sure about where we will spend eternity, in heaven or in hell, to think more about how our works will be assessed at the Judgment Seat of Christ, than to be obsessed with climate change. But it's not as simple as that! The Lord Jesus highlighted the importance of "loving our neighbour as ourselves", and the Bible stresses over and over again that as Christians we should be concerned about others. Indeed, by our good works which others see and should benefit from, we glorify our Father in heaven.[17]

Individual Efforts

As individuals then, we have a duty and an obligation towards the benefit of others, to do good to all, especially but not exclusively to "those who are of the household of faith".[18] How we do this will be an individual matter, but sharing what we have with others, preventing wastage of fuel, food and other

---

17 Matthew 5.16
18 Galatians 6.10

things, co-operating in efforts to improve air quality and land use can all come into the mix in the places where we live. Also complying as much as we can with government initiatives and regulations regarding waste and environmental issues should be seen as normal for a Christian. And we must not forget the clear instruction of Scripture to pray for governments and their officials, the "higher powers", the "minister of God to thee for good", and to render to each their due.[19]

No one of us can alter the big picture worldwide, nationally or internationally, but each of us can consider carefully how we use the resources God has provided us with. I speak generally, for I know this does not apply to everyone, but perhaps in our more affluent society many of us do need to move towards a less extravagant lifestyle, for example reappraising how many and what size of vehicles we own and drive, what travelling we do and why, what we spend on ourselves and on our homes when they are not needing it.

We can and should show that we care about others in our own workplaces and neighbourhoods. We can and should show that we care about those who live in countries where drought and scarcity already exist and are likely to get worse as global temperatures rise. We do not need to get agitated about "green" issues, nor be obsessed by our "carbon footprint". We need not get involved in politics, green or otherwise, nor in demonstrations and protests, peaceful or strident! But by being careful and caring we can commend the gospel to those around us, who watch us and expect much from us although they may not want to listen to the great message of grace we bring to them. As we do this, we will be representing Christ more consistently.

As Christians we should also be model citizens who are respected by those who know us, who comply with our government's

---

19 1 Timothy 2.1-2; Romans 13.1-7 (KJV)

legislation about all matters of common good, including these environmental matters and concerns. So wouldn't it be a good thing to practise the 3 Rs of environmental care – *Reduce, Reuse, Recycle* – and show to our neighbours and to others that we care? It's God's creation, after all!

National Efforts

After quite a slow start, many different nations and their governments have now taken more seriously the challenge of global warming and climate change. This may have been driven by self interest or a desire for international influence and reputation, a few even motivated by altruism and good will. Some see it as an opportunity to develop new technologies and improve employment opportunities, while others are more reluctant to abandon traditional industries upon which their prosperity and industry have depended for a long time. It is not surprising that individual nations will always try to safeguard their own economic interests first. Countries which have lagged behind, "third world countries" as they are often called, seem determined to catch up and at the same time seek compensation from richer countries for exploitation over previous centuries.

We must not lose sight of the fact that national boundaries do not apply to the atmosphere. The air we breathe is not localised, nor are our emissions into it, so that what happens in the east or the west or the north or the south affects everyone in between. Thus the efforts of some smaller countries whose emissions are already quite small can be cancelled out by what other larger countries with already huge emissions of CO_2 are doing and planning.[20] Smaller countries like the UK are vigorously developing wind farms on land and sea, keen on

---

20 Of the world total in 2020 - China, 32.5%; USA 12.6%; UK 0.9%

becoming "carbon neutral" within a short time, but they share the planet and its atmosphere with larger countries which are opening new coalmines to serve their power stations for decades to come. Localised, the effort seems to be a good thing, but in the context of worldwide significance, it can all seem to be overshadowed, even futile.

International Efforts

Up till recently there have been many difficulties in reaching international agreements even on targets and timetables for progress towards minimum CO_2 emissions. The most promising one was probably the United Nations Climate Change Conference held in Glasgow in November 2021.

COP26, as it was called, was attended by 25,000 delegates from 197 countries.[21] Previous conferences had set targets such as the *Paris Agreement* in 2015, reviewed annually. This time the *Glasgow Climate Pact* was produced which "reaffirms the Paris Agreement temperature goal of holding the increase in the global average temperature to well below 2 °C above pre-industrial levels and pursuing efforts to limit the temperature increase to 1.5 °C above pre-industrial levels ... [it] requires rapid, deep and sustained reductions in global greenhouse gas emissions, including reducing global carbon dioxide emissions by 45 per cent by 2030 relative to the 2010 level and to net zero around midcentury". However, achieving the target was not ensured, as even with existing pledges the emissions in the year 2030 will be 14% higher than in 2010. Proposals to drastically reduce the use of coal were also agreed, but "phased down" rather than "phased out". Many protesters believed that there was abundance of discussion and plenty good proposals but a lack of commitment and urgency. Cynics would say that there

21 COP 27 was held in Sharm-el Sheik in Egypt in November 2022

Could resolution of this problem be on the agenda of some future world leader?"

was plenty talk, but much less real, worldwide cooperation to tackle the problem - little urgency for change particularly by the countries where pollution is worst.

As an aside, the conference was described as receiving "the cleanest electricity in the UK", with 70% supplied from nuclear power from Torness and Hunterston B, the rest from wind power (I'm not sure how they measured and guaranteed this!). Also, councils in and around Glasgow pledged to plant 18 million trees during the following decade.

It is obvious that to be effective, worldwide cooperation is necessary if global warming is to be controlled and the worst effects of climate change avoided. But this has not been easy to achieve because individual nations want to defend and safeguard their own economic interests and, for some, their political bias is not helpful.

Could resolution of this problem be on the agenda of some future world leader? The Bible tells us that such a one is due to arrive on the scene, perhaps sometime soon, with a radical agenda![22]

To think more about

Do you think these targets for CO_2 reductions and for temperature rise limited to 1.5 °C are achievable?

Do you think you could do more to limit your own CO_2 emissions? How? Would it be worth the effort?

What should a Christian's emphasis be in daily living?

---

22 See 2 Thessalonians 2 and Revelation 13

CHAPTER 8
CLIMATE CHANGE IN THE PAST

Geologists and earth scientists tell us that there is good evidence that in the distant past, the earth has undergone several changes to its climate. They tell us that ice ages have come and gone when worldwide temperatures were extremely low. At other times the climate was uniformly benign, and water created great changes to the landscape and the seas. There may have been meteor strikes to create great craters in some places, and fossils tell their story.

More recent historical and anecdotal evidence points to more recent times of extreme cold. The River Thames froze over 23 times between 1600 and 1814, a period which has been called the "Little Ice Age". People walked and skated on it, and "Frost Fairs" were set up in tents and booths pitched on the ice, with all kinds of commerce and sports being held, some of them rather notorious. The coldest winter was 1683-84 when even the sea at the south coast froze solid up to 2 miles from the shore. The last time the Thames froze over was during the very cold winter of 1963, when some scientists were supporting a 'Global Cooling' hypothesis! There have also been some very wet and also very dry seasons in the UK.

We may be sure, however, that the biggest ever climate change on Earth was caused by the great flood documented clearly in the Bible in Genesis 6-8. All the effects mentioned in the first paragraph of this chapter can be explained more cogently and logically with reference to that one flood and its aftereffects which certainly changed the climate and the landscape too. Natural, materialist explanations of the evidence are different from those of Bible-believing Christians, who read in 2 Peter 3.5-6, "For this they willfully forget: that by the word of God the heavens were of old, and the earth standing out of the water and in the water, by which the world that then existed perished, being overflowed with water". (Note the word 'willfully' in the first sentence!) The world today is much different from what God made in the beginning. Present day geography and topography does not describe the appearance of the world before the flood. The climate changed and so did much else.

The Flood

So let's remind ourselves of the great flood and its effects, and remember too how it is referred to in the Old Testament outside of the Genesis account, and how in the New Testament its spiritual significance is highlighted.[23] Here is a brief description of the flood and its effects - a more detailed account is available elsewhere.[24]

When the flood came, truly vast amounts of water fell from the skies for forty days and nights without interruption. The waters which God had put "above the firmament" on Day 2 of creation for protection and blessing were now used for destruction and judgment as they deluged and inundated the earth. In addition, when we read that "all the fountains of the great deep were broken up" (Genesis 7.11), it could signify intense volcanic activity from deep below the earth's crust which would break

23 Job 22.15-16; Isaiah 54.9; Matthew 24.37-39; Hebrews 11.7; 1 Peter 3.20; 2 Peter 2.5
24 *Creation's Story,* R W Cargill, pg 56-70, John Ritchie Ltd, 2008

down shore barriers and allow sea water to travel inland. It would also release molten material, lava, solidifying on contact with water and generating clouds of steam. An alternative view is that these "fountains" were subterranean waters surging upwards, all this to give a flood of cataclysmic dimensions which we can hardly imagine. Whatever it means, the damage and change caused by deepening surface water, huge swirling sediments, volcanic activity and much else must have been immense. It is obvious that the climate would be changed radically, and the landscape too, away beyond what it was like in antediluvian times.

Fossils

Buried in the sediments and quickly compressed, huge numbers of creatures would be fossilised. As the great flood advanced, the first creatures to be buried would be the least mobile, those which lived in the shallow seas and around the shore line. The more mobile ones would seek refuge higher up until overtaken by the flood, and the last to succumb would be the most mobile animals and birds at the highest levels.

The 'geological column' is widely taught and believed without question to be a historical record of how creatures evolved over millions of years. It is not. It is rather a record of where and when they lived and died when the flood came and destroyed them all. The order of their burial and fossilisation is according to their habitat - aquatic life first, the bottom dwellers like shells and trilobites, followed by the fishes in huge shoals buried in the contorted swimming positions which their fossils show. Then the shoreline amphibians and the slow reptiles are overcome and buried, and eventually the mammals, often in huge numbers as they herded together in the race to escape, as testified by vast fossil beds worldwide. All this fits the observed geological fossil record.

Antediluvian Climate

It appears that rainfall as such was unknown before the flood. Plant growth and other water needs would be provided for by a cyclical "mist" which rose from the earth's surface by solar heating and returned possibly at night (see Genesis 2.5-6). Furthermore, that large amount of water which was above the aerial atmosphere ("firmament") in vapour form would have an important effect on the earth's climate and on the well-being of every one of its inhabitants. Firstly, it would give a positive "greenhouse effect". It would prevent extremes of heat and cold, so that a uniformly temperate or subtropical climate would have existed all over the world, giving conditions for luxuriant growth of vegetation and effective reproduction of animals and, of course, man. Secondly, it would filter out most of the harmful cosmic rays which bombard Earth continuously, and also remove much harmful ultraviolet radiation, shielding everything from these particles and rays which are known to cause damage to living tissue and produce disease and ageing. So animals and plants lived longer and grew bigger - as fossilised samples show.

So, in antediluvian times, although the peace and beauty of Eden itself had gone due to the Fall, climate conditions were benign, ideal for life and growth. It is little wonder, therefore, that people lived so long as recorded in the early chapters of Genesis, and had large families. In addition, animal and plant life would thrive, so that huge animals, e.g., mammoths and dinosaurs in all their variety could exist all over the world, and huge forests, grasslands, and swamps of immense variety and density would be normal, with huge specimens in them. When buried and compacted during the flood, all this vegetation would provide the world's enormous coal beds, and by a different mechanism involving different organisms, vast oil fields also.

A Changed World

After the flood it was all so different across the globe. Radical climate change had occurred. Without the shielding of the high altitude vapour canopy, extremes of temperature and contrasting seasons would become the norm (see Genesis 8.22, the first references in the Bible to summer and winter, cold and heat). With the drying up of the flood, mighty winds blew (8.1), surface waters would freeze, ice sheets would form and make their marks on the landscape as they moved and melted. Mankind would face a new set of difficulties which would have far-reaching effects. And many plant and animal species which had been common before the flood could exist no longer, unable to adapt to the changed environment, except for a few which found a niche in which to survive. This is when 'survival of the fittest' would really apply, and is mainly why dinosaurs and similar land giants would become extinct.

The world and its climate were very different before and after the flood. Indeed, that is what 2 Peter 3.5-7 tells us, as we have noted before. This was truly dramatic, indeed drastic, climate change. It happened suddenly, thousands of years ago, and brought about changes which have affected mankind to this day, and have affected the whole natural world and its wildlife too.

Before leaving this chapter, let's look again at 2 Peter 3. 7: "But the heavens and the earth which are now preserved by the same word, are reserved for fire until the day of judgment and perdition of ungodly men". And then verse 10: "The day of the Lord will come as a thief in the night, in which the heavens will pass away with a great noise, and the elements will melt with fervent heat; both the earth and the works that are in it will be burned up". This is something much bigger for a future time, not just climate change but a greater change in the very fabric of the earth. We'll look at this briefly in our next chapter.

But note the connection of this future judgment with the past one which we have just described. In Matthew 24.37-39 the Lord Jesus linked them together. Just as the first world was destroyed by water, the present one will be destroyed by fire, both being divine judgment on unabated human sin. The world after the flood lost much of the glory and beauty associated with its first creation, but after the fires of judgment have passed, the world to come will be one of greater glory and beauty. There will be a new heaven and a new earth where sin and evil have been banished forever, all war and want, all misery, disease and death gone, never to return.

The coming of our Lord Jesus Christ at the Rapture, and His subsequent appearing in glory, will bring about the greatest and best change ever! And Christians can say confidently, "We shall be changed ..."

We will not only be with Christ but "we shall be like Him, for we shall see Him as He is".[25]

To think more about

The Genesis records of Creation and then the Flood have been greatly disputed and even doubted and ridiculed. Why do you think this is?

And why should Christians take these records literally and completely?

---

25 1 John 3.2

CHAPTER 9
CLIMATE CHANGE IN THE FUTURE

We have noted how the great Flood as described in Genesis 6-8 had the biggest ever effect on the world's climate, changing dramatically the atmosphere, the surface of the planet and its living things. Turning now from the first Book in the Bible to the last one, we discover that much greater and more severe climatic events will overtake this sad world in the future before it is completely transformed under the kingship of our Lord Jesus Christ.

In some other parts of the New Testament there are definitive statements about future events in the world, events which are part of God's plan for purging evil from the earth and judging evil doers. As we go on to read John's visions as described in Revelation 6-19, we learn how such acts of divine wrath and judgment will come to pass, including how the evil Antichrist will be dealt with. As these judgments progress, they will involve great changes in the environment, changes beyond anything yet experienced, fearful to contemplate. Our Lord Jesus called it, "great tribulation".[26] Well might John the Baptist warn his hearers to "flee from the wrath to come" (Matthew 3.7), and how grateful we should be that by the grace of God, our Lord Jesus Christ is the one "who delivers us from the wrath to come" (1 Thessalonians 1.10).

---

26 Matthew 24.21

Seals, trumpets and bowls

These fearful judgments are described as the opening of the
seven seals of a book by the Lord Jesus Himself, "the Lamb"
(Revelation 6), then the sounding of seven trumpets by seven
different angels (chapters 8-9), and finally another seven angels
pouring out upon the earth seven bowls (vials) filled with the
wrath of God (chapter 16). They involve cosmic and terrestrial
events which shake the world to its core, bringing destruction
to much of the environment, in addition to widespread disease
and loss of life on a huge scale.

As we read about these fearful judgments and how they were
brought about, it is not difficult to see how at least some of
them will involve drastic climate change, and by means of
which many millions of people on earth will be killed. While it
is inappropriate to speculate, and unwise to be dogmatic, it is
reasonable to believe that natural and cosmic forces might well
be involved in these events. The following list is not complete,
but it illustrates this point.

- *Seal No 6* (ch 6.12-14): A great earthquake moves
  mountains and islands, the sky blackens and stars
  (meteors?) fall to the earth.

- *Trumpets Nos 1 -3* (ch 8.7-11): Fierce storms of lightning
  destroy a third of the earth; a burning meteor or comet
  falls into the sea and destroys a third of it; another similar
  burning "great star" called Wormwood poisons a third of
  rivers and fountains.

- *Trumpet No 4* (ch 8.12): the light of the sun and moon
  and stars is obliterated for a third part of the day and of
  the night.

- *Bowl No 2* (ch 16.3): the sea becomes blood and all sea
  life dies.

For long centuries the earth has been mismanaged and plundered."

- *Bowl No 3* (ch16.4): all rivers and fountains of water became blood.

- *Bowl No 4* (ch 16.8): the sun's heat becomes unbearably scorching.

- *Bowl No 5* (ch 16.10): darkness envelops the kingdom of "the beast".

- *Bowl No 7* (ch 16.17-18): the greatest earthquake ever occurs along with thunders and lightning.

1,000 Years of Peace

One final transformation of the earth and its climate will take place after all these fearful judgments have done their work. The climate will change from being destructive, as we have just noted, to being benign and beautiful, perhaps very similar to what it was in Eden at the very beginning. Christ will return to earth in power and glory as the King of kings and Lord of lords, defeating every enemy of God and man. He will destroy that Man of Sin, the Antichrist, and judge the confederate nations of this sin-blighted earth. Then He will introduce His kingdom of righteousness, peace and joy in a renewed and stable earth as promised and prophesied many times in the Old and New Testaments. It will last for 1,000 years. 'Paradise Lost' will at last become 'Paradise Regained'.[27]

For long centuries the earth has been mismanaged and plundered. Its resources have been wasted. Its real Owner has been set aside and ownership of sections of it has been hotly

---

27 John Milton, 1667

and repeatedly disputed. Different nations and peoples have attacked and cruelly exploited others in wars and conflicts which have produced immense suffering for millions.

All this will cease when our Lord Jesus Christ reigns in righteousness and justice. At last there will be an ideal time of peace and plenty in the wonderfully changed climate of an earth renewed and restored by the power of God.

The Old Testament contains many descriptions of this wonderful millennial kingdom. For example, the prophet Isaiah described how different it will all be. Here are some of the things that will **not** happen then.

- *There will be no wastage agriculturally* - the climate will be better for plants to grow (30.23-26). Pests and diseases will no longer cause problems. Supplies will be freely available to everyone so that hunger and thirst will not exist anywhere.

- *There will be no hurting environmentally* – the nature of animals will be changed (11.4-9 and 65.25). Conditions will become like what they were at the beginning of time before sin entered the world.

- *There will be no fighting internationally* – the world will be free from war (2.2-5). All the huge expense of military defence and warfare will be used for beneficial and peaceful purposes.

- *There will be no crying personally* - disease and death will not prevail (25.6-8). Expensive health care and medical intervention will not be needed.

All this will be possible because "a King will reign in righteousness" (32.1). The Prince of Peace (9.6) will be in control. "The earth shall be full of the knowledge of the LORD as the waters cover the sea" (11.9).

There will be positive change –

> Spiritual Change,
> Moral Change,
> Governmental Change,
> Physical Change,
> Climate Change!

What a wonderful time that will be! For so long the earth has been mismanaged and plundered, its peoples exploited, its resources wasted, its ownership disputed. Our Lord Jesus Christ will reign in righteousness and justice, and a Eutopian time of peace and plenty will exist at last. Everyone under His benign rule will live without fear of the effects of global warming, wars, pandemics, disasters, or any such thing.

To think more about

Can you connect the words of the Lord Jesus in Matthew 24 to the visions of John in Revelation 6, for example?

What amazes you most about the future – on earth and in heaven?

CHAPTER 10
TO SUM UP BRIEFLY

- Worldwide climate is changing, being driven by a steady global warming. The extra heating is due to excess carbon dioxide in the atmosphere mostly coming from combustion of carbon based fuels.

- Extra heating evaporates more moisture (water vapour) from the earth's surface giving rise to droughts, wild-fires, and famines. The extra water in the clouds gives rise to heavier rainfall and more flooding.

- Most governments in the world have recognised the problems and their future implications. United Nations is attempting to obtain international agreements in a unified approach to reducing emissions.

- However well meaning these efforts are, they are based on a totally materialistic view of the world and its future, without any reference to God and His revelation.

- Christians know that this world and its climate have been changed before, and are due to be seriously damaged during a period of around seven years in what is called, 'The Great Tribulation', a fearful time when God will judge unrepentant mankind for its persistent sin.

- This will be followed by a period of unprecedented peace and justice when our Lord Jesus Christ rules over all the earth for 1,000 years with a climate more benign than ever.

- Eventually, the world and its total environment will be annihilated ("burned up") and God will introduce a new heaven and a new earth in which righteousness will dwell forever.

- Such changes to planet Earth are certain in the future, but that is not a reason for Christians to ignore the present implications of climate change and adopt a careless and extravagant lifestyle. Christians can show to others that they care, and cooperate with governments in well-meaning efforts to protect the environment.

- We need not get agitated about "green" issues, nor get involved in politics and demonstrations, but we should do all we can to represent Christ and commend the gospel by how we live.

- The whole environment is God's creation, and man was entrusted with its stewardship. Sadly neglected and abused in man's quest for more power and wealth, much has suffered. No one of us can alter the big picture, but each of us can consider carefully how we use the resources God has provided.

- Heaven is our true home for all eternity. Spiritual and eternal matters need more of our attention than physical and temporal ones. But as Christians we surely cannot overlook or neglect matters which affect the welfare of our fellow human beings.

INDEX

More books in the series ...

The Money Maze - The Christian and Money
Stephen Baker

Money has a strange effect on people. For some people, accumulating money and creating wealth is a habit, and the aim of their life is to ensure that they are never without it. Others view money as a means to an end – necessary, but in itself of no value - its only purpose being to help them get through life with as little hassle as possible. There are people who consider money to be evil in itself, as it brings out the raw, sinful responses of greed, avarice, covetousness and jealousy. At the opposite end of the spectrum, there are those who see money as a positive thing to be desired, and a visible evidence of success, skill, creativity and intelligence. This book explores the truth about what money is, and looks at it from a Biblical perspective. In other words, what does the Bible say about money?

ISBN: 9781914273124 John Ritchie Ltd., (2021)
www.ritchiechristianmedia.co.uk

Made in His Image - The Christian and Abortion
Dr Philip Mullan

The abortion debate raises sensitive issues, and often provokes strong reactions on both sides of the argument. This short book addresses these issues in a holistic and Bible-based manner, by exploring more than just the hard facts of abortion. We are reminded of the Bible's teaching on what it means to be human, and of the high value that God places on every individual, including the unborn child, the teenage mum, and those who have had abortions. The question of when human life begins is explored from both a biblical and a scientific perspective. Finally, the real heartache, distress and fear that many women and couples face is considered, and we are encouraged in ways that we can genuinely support them and their children.

"Some may say that abortion is a woman's issue, and that, as a man, I should not speak about it. But, as a Christian, I believe God has spoken about it, and His voice must be heard. We must be willing to submit to His truth, knowing that He is good, and His ways are best." (Dr Philip Mullan).

ISBN: 9781914273131 John Ritchie Ltd., (2021)
www.ritchiechristianmedia.co.uk

Connecting Church & School - Teaching the Bible to School Children
Paul Coxall

In a world full of different philosophies and ideologies, it is vitally important that the message of the Bible, a book that has transformed the lives of many, as well as defining the laws and culture of numerous countries, is made known.

Many years ago, a servant of God said "Here I am! Send me" (Isaiah 6:8b ESV). Today there is a great need for Believers to echo that same statement and to teach young children the wonderful message found in the Bible.

This book has been written, not by an expert, but someone who has endeavoured, through the local church, to develop a work in local schools. The author has drawn on his experience and observations to offer guidance and advice to anyone who has a desire to serve the Lord, by helping young children understand what the Bible teaches. Although the book is principally aimed at connecting with and teaching in Primary Schools (Children aged from 5-11 years old), it does contain useful lessons for those looking to teach older children. It will be a helpful guide, and offer encouragement to all who desire to work with local schools and teach children the transforming truths found in God's Word.

ISBN: 9781914273377 John Ritchie Ltd., (2023)
www.ritchiechristianmedia.co.uk